# What is...
# ?

# *Bendy*

**Heinemann**

First published in Great Britain by Heinemann Library
an imprint of Heinemann Publishers (Oxford) Ltd
Halley Court, Jordan Hill, Oxford OX2 8EJ

MADRID ATHENS PARIS
FLORENCE PRAGUE WARSAW
PORTSMOUTH NH CHICAGO SAO PAULO
SINGAPORE TOKYO MELBOURNE AUCKLAND
IBADAN GABORONE JOHANNESBURG

© Heinemann Publishers (Oxford) Ltd

Designed by Heinemann Publishers (Oxford) Ltd
Illustrations by Samantha Elmhurst
Printed in China

99 98 97 96 95
10 9 8 7 6 5 4 3 2 1

ISBN 0431 07979 X

**British Library Cataloguing in Publication Data**
Warbrick, Sarah
Bendy. - (What is...? Series)
I. Series
500

Acknowledgements
The Publishers would like to thank the following
for the kind loan of equipment and materials used in
this book: Early Learning, Bishop Stortford. Toys Я Us Ltd,
the world's biggest toy megastore.

Special thanks to Bryan, George, Jodie, Joel and Michael
who appear in the photographs

Photographs: Action Plus pp14–15; Bruce Coleman p8;
S&O Matthews pp18, 19; TSW p10; other photographs by Trevor Clifford
Commissioned photography arranged by Hilary Fletcher
Cover Photography: Trevor Clifford

There are bendy things all around us.
Bendy things can be fun.
Bendy things can be useful.
Be careful, if things bend too far,
they can break!

This book shows you what is bendy.

These things look different.
What differences can you see?

In one way they are the same.
They are bendy.

3

This train is made of
carriages linked together.

It bends around the track.

This bridge is made from
lots of pieces of wood.

Look how it bends when you
stand on it.

The cat's tail is bendy.

This is a skeleton of a cat's tail.
Look at the small bones, linked together.

How can this ballet dancer bend so
far backwards?

The bones in her back are linked to
help her bend.

Look at all this liquorice.

It is very bendy.
Bryan can make all sorts of
different shapes.

The pole vaulter is using a long pole to help him jump very high.

The pole bends as he flies into the air.

Look at these tent poles.
They look strong and straight.

But they bend easily to
help Michael put up the tent.

Do you think this metal bar can bend?

When the blacksmith heats it in
the fire, it will bend.

What is bendy here?

# Index